BASIC / NOT BORING

MAP SKILLS & GEOGRAPHY

Grades 2-3

Inventive Exercises to Sharpen Skills and Raise Achievement

Series Concept & Development
by Imogene Forte & Marjorie Frank
Exercises by Marjorie Frank

Incentive Publications, Inc.
Nashville, Tennessee

Thanks to Maurine Bridges for her assistance

About the cover:
Bound resist, or tie dye, is the most ancient known method of fabric surface design. The brilliance of the basic tie dye design on this cover reflects the possibilities that emerge from the mastery of basic skills.

Illustrated by Kathleen Bullock
Cover art by Mary Patricia Deprez, dba Tye Dye Mary®
Cover design by Marta Drayton, Joe Shibley, and W. Paul Nance
Edited by Anna Quinn

ISBN 0-86530-397-5

2   3   4   5   6   7   8   9   10   08   07   06   05

PRINTED IN THE UNITED STATES OF AMERICA
www.incentivepublications.com

TABLE OF CONTENTS

Appendix

CELEBRATE BASIC SOCIAL STUDIES SKILLS

Basic does not mean boring! There is certainly nothing dull about . . .
  . . . following huge footsteps to Bigfoot's cave
  . . . trying to deliver mail in a crocodile-infested swamp
  . . . figuring out where to deliver a pizza, a lobster, or a newborn baby rat
  . . . searching for pirate's treasure or endangered African animals
  . . . delivering skeletons to schools or worms to a garden
  . . . helping a lost dinosaur or stagecoach driver read a map
  . . . seeing that heaters get delivered to the right igloos in the Arctic
  . . . finding a way to get flowers delivered to the queen bee in her hive

These are just some of the adventures students can explore as they celebrate basic social studies skills. The idea of celebrating the basics is just what it sounds like—enjoying and improving geography skills. Each page of this book invites young learners to try a high-interest, visually appealing geography or map exercise that will sharpen one specific skill. This is not just an ordinary fill-in-the-blanks way to learn. These exercises are fun and surprising. Students will do the useful work of practicing map and geography skills while they enjoy helping interesting creatures deliver unusual items all over the world.

The pages in this book can be used in many ways:
  • to review or practice a social studies skill with one student
  • to sharpen the skill with a small or large group
  • to start off a lesson on a particular skill
  • to assess how well a student has mastered a skill

Each page has directions that are written simply. It is intended that an adult should be available to help students read the information on the page, if needed. It is also important to have maps and geography resources available to students as they do the exercises. In most cases, the pages will be used best as a follow-up to a skill that has already been taught. The pages are excellent tools for immediate reinforcement of a skill or concept.

As your students take on the challenges of these adventures with maps and geography, they will grow! And as you watch them check off the basic social studies skills they've strengthened, you can celebrate with them.

The Skills Test

Use the skills test beginning on page 57 as a pretest and/or a post-test. This will help you check the students' mastery of basic social studies skills and prepare them for success on achievement tests.

SKILLS CHECKLIST
MAP SKILLS & GEOGRAPHY, GRADES 2-3

✔	SKILL	PAGE(S)
	Locate own home in the universe	10–11
	Locate own country, home, and neighboring countries on a world map	10–11, 33
	Place items on a map	11, 13, 22, 29, 44–46, 56
	Use maps to locate things and places	11–24, 30–40, 48–52, 54–55
	Define a map as a representation of a real place	12–13
	Identify and use a variety of maps	12–28, 30–40, 48–52
	Identify the parts of a map (title, labels, scale, key, compass)	14
	Choose a title for a map	15
	Use directions to locate things on a map	16–19, 54-55
	Identify directions on a map	16–19, 54-55
	Compare locations on a map	17–19, 22, 36–40, 51, 54–55
	Use maps to find information and answer questions	17–24, 26–28, 38–52, 54–55
	Identify and draw map symbols	20–22, 25
	Use map symbols to find things on a map; match symbols to words	20–25
	Read a map key and use it to locate things on a map	20–25
	Make a map key	25
	Use a simple scale to determine distances on a map	26–28
	Recognize landforms on a map	29
	Identify poles, equator, and hemispheres on a map	30–31
	Recognize and identify different kinds of maps	30–31, 48–52
	Recognize continents and oceans on a world map	30–33
	Recognize boundaries on a map	33
	Identify some states on a U.S. map	35–37, 40
	Recognize names and shapes of some states in the U.S.	35–37, 40
	Identify and locate some cities and landmarks in the U.S.	36–40
	Locate own state on a U.S. map	37
	Identify some major bodies of water and cities in the U.S.	38–39
	Find information on charts	41, 42
	Find the location of objects located on a simple grid	43–47
	Follow directions to place objects on a grid	44–46
	Find information on a population map	48
	Find information on an environmental map	49
	Find information on a product map	50
	Find information on a rainfall map	51
	Find information on a road map	52
	Recognize a variety of geographic terms	53
	Make simple maps	56

MAP SKILLS
& GEOGRAPHY
Grades 2-3

Skills Exercises

Where Are You in the Universe?

You've got mail! This mail delivery creature from outer space has a letter for you.

She needs your address in the universe!

Fill in your universe address, and then help her find you by following the other directions.

COSMOS

From Outer Space

TO:

Name _____

Street _____

City, State, or Province _____

Country _____

Continent _____

Planet _____

Galaxy _____

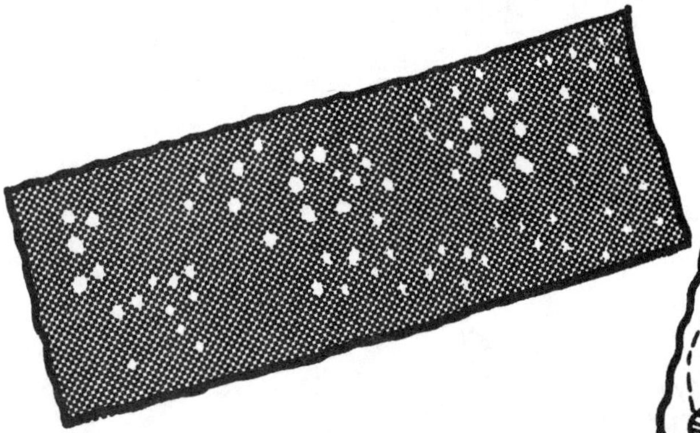

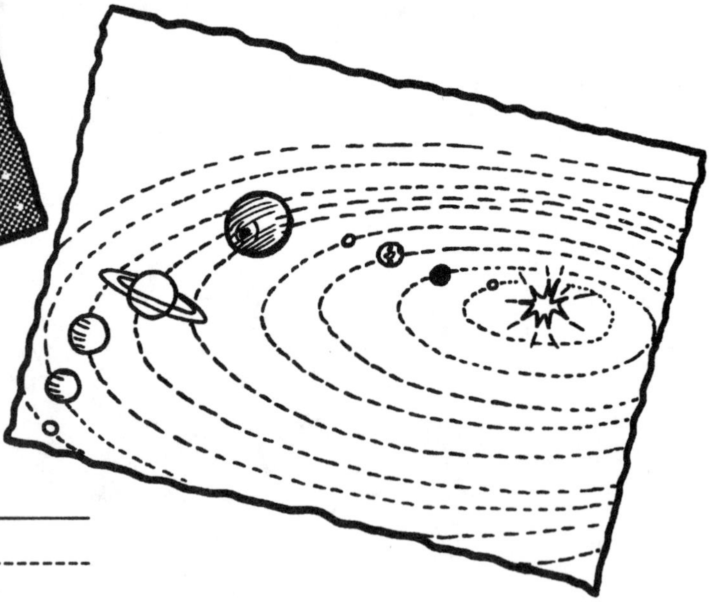

Write the name of your galaxy.

_____
- -
_____

Color in your planet on the solar system.

Name _____

Where Are You in the Universe? cont.

Color your continent. Put a star where you live.

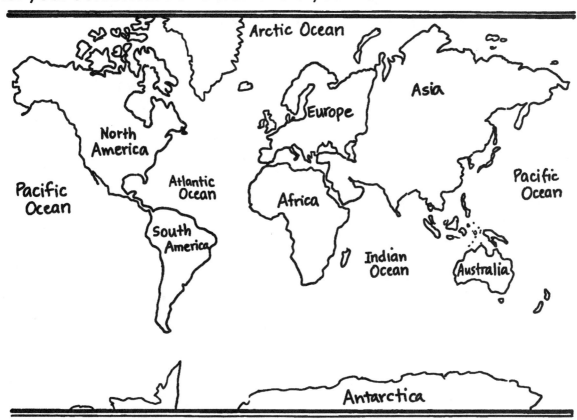

Draw a picture of your home.

Name _____

Use with page 10.

⑪

Once upon a Table

It's not easy delivering ice cream to desert folks on a hot summer day.
Harry needs a big breakfast before starting his busy day.

A map is a flat picture of a real place.
Look carefully at the table. Then look at the maps of the table.
Which map matches Harry's table?
Color the map that is correct.

Now make a map of
Harry's ice cream wagon.

See the picture on the
next page (page 13).

Name _____

Harry's ice cream delivery wagon is loaded and ready to go.
Make a map of his wagon for him, so he will know where everything belongs!

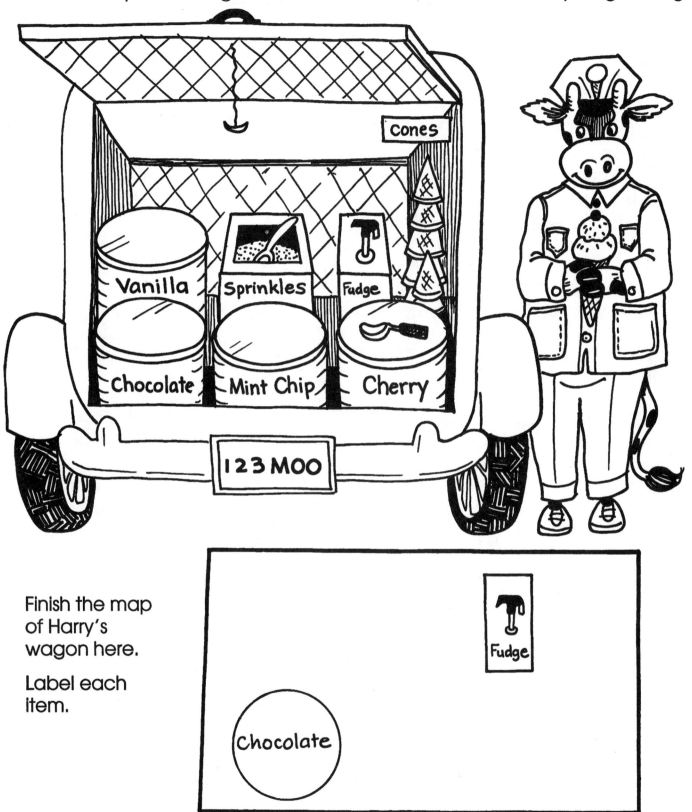

Cones

Vanilla   Sprinkles   Fudge

Chocolate   Mint Chip   Cherry

123 MOO

Finish the map
of Harry's
wagon here.

Label each
item.

Fudge

Chocolate

Name _____

Digging for Treasure

Arvin's Airline is bringing treasure hunters to Skull Island to dig for treasure.
They have found an old treasure map.
Now they have to decide what it means.

Read about each part of the map. Follow the directions.

The **title** tells what the map is about. Draw a red circle around it.

The **map key** explains symbols and pictures that stand for things on the map. Color it yellow.

The **scale** tells about distances on the map. Draw a green box around it.

The **compass** shows directions. Color it orange.

The **labels** tell what things on the map are. Draw a purple circle around each label.

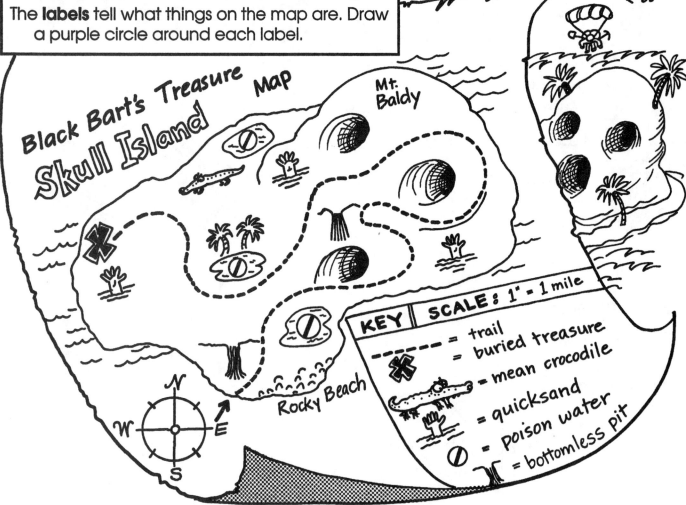

Name _____

Map Mix-up

Carl Crane has a package to deliver on Maple Street.
The titles have fallen off his maps, and he doesn't know which one to use.
Choose the best title for each map. Circle it.
Then color the map that Carl needs to make his delivery.

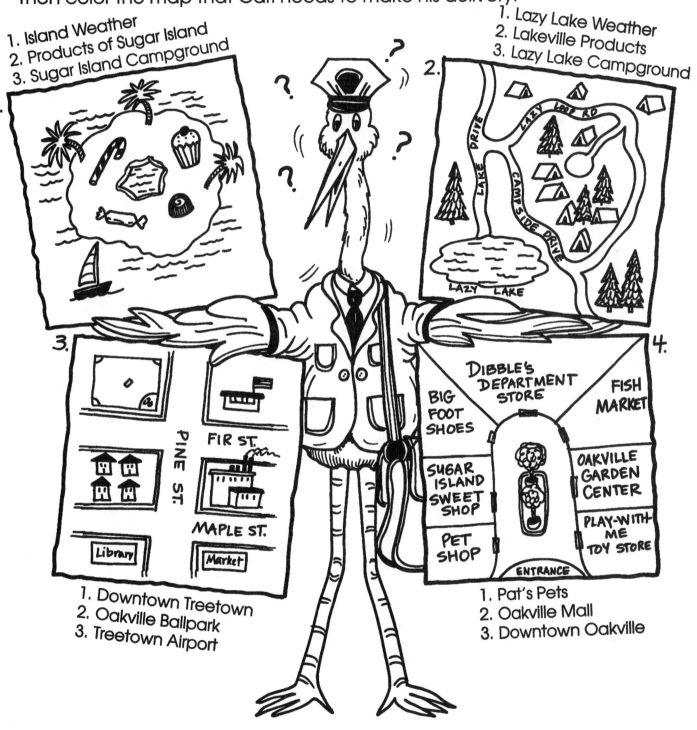

1.
1. Island Weather
2. Products of Sugar Island
3. Sugar Island Campground

2.
1. Lazy Lake Weather
2. Lakeville Products
3. Lazy Lake Campground

3.
1. Downtown Treetown
2. Oakville Ballpark
3. Treetown Airport

4.
1. Pat's Pets
2. Oakville Mall
3. Downtown Oakville

Name _____

Delivery for Bigfoot

Mail Carrier Millie has a delivery for Bigfoot.
Help her follow the footsteps to Bigfoot's cave.
Follow the directions.
Color a path along the footsteps as you go.

1. Walk 4 steps south.
2. Walk 2 steps west.
3. Walk 4 steps south.
4. Walk 6 steps east.
5. Walk 2 steps north.
6. Walk 5 steps east.
7. Walk 5 steps south.
8. Walk 1 step west.
9. Walk 2 steps south.
10. Walk 1 step east.

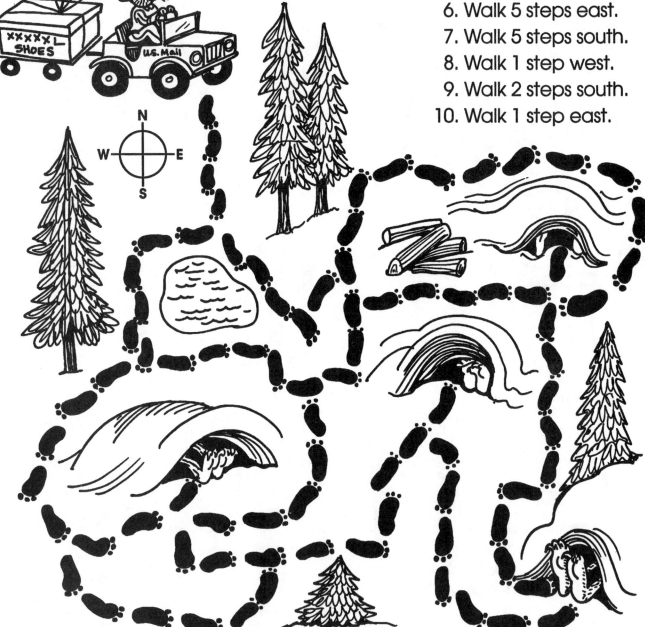

If you followed the directions correctly, you should be there!

Name _____

A Dangerous Delivery

Francie Frog always dreads bringing mail to the swamp.
She is very careful when she reaches out to a mailbox!
Today she has several packages to deliver.

Help her find the right
mailboxes.

Write **N, S, E, W,**
NE, SE, NW, or **SW** in each blank.

1. Turtle's box is _____ of Toad's.

2. Crayfish's box is _____ of Lizard's.

3. Toad's box is 1 box _____ of Pelican's.

4. Rat's box is _____ of Crocodile's.

5. Possum's box is _____ of Crayfish's.

Write the answer in each blank.

6. _____ box is just south of Toad's.

7. Crayfish's box is west of _____ .

8. _____ box is northwest of Crocodile's.

9. Is Pelican's north of Possum's? _____

10. Is Rat's box a little northeast of Lizard's? _____

11. Whose box is farthest north? _____

Name _____

Basic Skills/Map Skills & Geography 2-3

Directions on a Map

Delivery by Dinosaur

Where is the manager of the Fossil Fun Park?
The mailman needs the manager's signature for a special package.
Follow the directions to help him find his way to the
manager's office. Draw a line to show the path he should take.

1. Go 3 steps south.   5. Go 6 steps east.
2. Go 10 steps east.   6. Go 1 step south.
3. Go 6 steps south.   7. Go 3 steps east.
4. Go 8 steps southwest.

MAIL POUCH

ENTER

Bucking Stegosaurus Ride

Fossil Hill

Pterodactyl Hop

See-saw-o-saurus Ride

Snack Bar

NW N NE
W   E
SW S SE

CLOSED-NAPTIME
MANAGER'S OFFICE

Name _____

Who Gets the Milk?

Here comes the milk truck to the Moptown Trailer Park!
Follow the directions to figure out who will get milk.

1. Only water is drunk in the trailer west of the milk truck. Color this trailer brown.
2. Milk is delivered to the trailer northeast of the truck. Color this trailer green.
3. Honey is the favorite drink in the trailer south of the truck. Color this trailer yellow.
4. The animal in the trailer southeast of the truck only drinks soda pop. Color this trailer orange.
5. The trailer north of the truck orders cactus juice. Draw a green box around it.
6. Milk is chosen by the animals to the northwest and southwest of the truck. Circle these two trailers, and then color them brown.

Name _____

Who Ordered the Pizza?

Popple, the pizza delivery girl, is on her way to deliver a double cheese, double pepperoni pizza to some hungry pizza-lovers in Rivertown.
Help her find her way around the city!

> Many maps have symbols to show where things are. Symbols are small signs or pictures that stand for large things on the map.

Look at the symbols on the map key on the next page (page 21).
The words tell what the symbols mean.
Follow the directions and answer the questions.
You'll also find out who ordered the pizza!

1. Is the pencil factory on Speedway Blvd.? _____

2. Trace the nature trail in red.

3. Color the playground green.

4. Is the shoe store in the same block as the grocery store? _____

5. Is the post office in the same block as the airport? _____

6. Is the movie theater next to the gas station? _____

7. What is across Bridge St. from the school? _____

8. What street is Poppie's Pizza on? _____

9. The church is on the corner of Speedway and _____ Blvd.

10. If Popple runs out of gas by the church, how many blocks will she have to walk to get to the gas station? _____

11. The pizza is delivered to workers at the building on the southwest corner of Speedway Blvd. and North Ave. Who ordered the pizza? _____

Name _____

Use with the map on page 21.

Map Symbols

Basic Skills/Map Skills & Geography 2-3

= airport		🛒 = grocery store			
✏️ = pencil factory		👟 = shoe store			
⛪ = church		🛝 = playground			
📖 = library		= picnic area			
🏫 = school		⛽ = gas station			
➕ = hospital		🎥 = movie theater			
✉️ = post office		⊙ = Poppie's Pizza Palace			

= street
= railroad track
= bridge
= nature trail

Map of Rivertown

NORTH AVE.

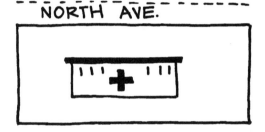

DOWNTOWN BLVD.

SPEEDWAY BLVD.

RIVER WAY

BRIDGE ST.

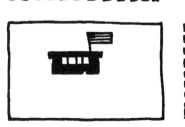

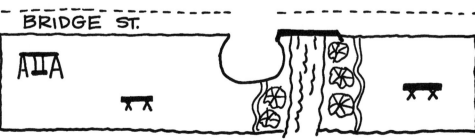

Use with page 20.

㉑

Map in a Candy Box

Curtis, the candy delivery guy, has a large box of candy for Miss Pansy.

Oops! He spilled all the candy!

Luckily, the lid of the candy box has a map key that tells where each kind of candy belongs.

Use the key to put the candy back in the box before Miss Pansy opens the door!

Draw and label the correct candy in each space in the box.

OOPS!

Vanilla Cream

Mint Cream

Orange Taffy

Chocolate Drop	Butter-nut	Cookie Stick
Lemon Taffy	Vanilla Cream	Mint Cream
Cherry Cream	Nut Cluster	Orange Taffy

Chocolate Drop

Lemon Taffy

Cookie Stick

Cherry Cream

Nut Cluster

Butter-nut

Creamy-Dreamy Chocolates

Name _____

Basic Skills/Map Skills & Geography 2-3

A New Desert Route

Rod Roadrunner's mail route will take him across the desert every day.
Today is his first day of work, and he's learning to read his new map.
Help him get to know his map.

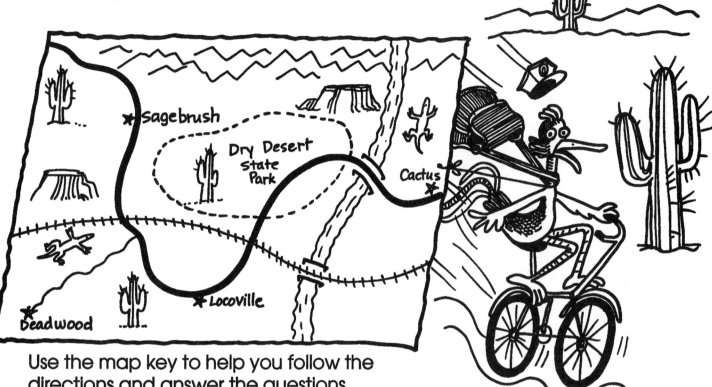

Use the map key to help you follow the
directions and answer the questions.

1. If Rod takes the highway,
   what is one city he will pass? _____

2. Color the mountains purple.

3. Color the cactuses green.

4. Color the dry riverbed brown.

5. How many lizard preserves are there? _____

6. How many plateaus will he see? _____
   Color them red.

7. How many times does his
   route cross the railroad? _____

8. What is the name of the park? _____

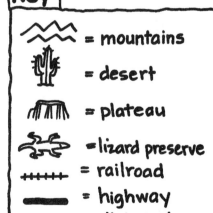

Key

∿∿∿	= mountains
🌵	= desert
⛰	= plateau
🦎	= lizard preserve
+++++	= railroad
——	= highway
～	= dirt road
≈≈≈	= dry riverbed
- - - - -	= state park
✶	= city

Name _____

Use a Map Key

Special Delivery

The stork has a special delivery for one of the families in the Critter Condos.

Follow the directions to find out where the new baby belongs.

Draw a line to follow the stork's path.

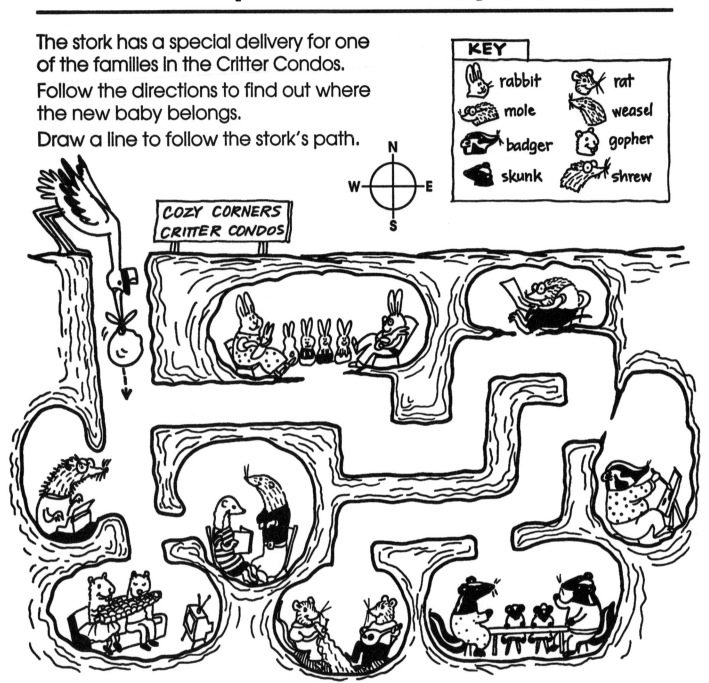

KEY
- rabbit
- mole
- badger
- skunk
- rat
- weasel
- gopher
- shrew

COZY CORNERS CRITTER CONDOS

Follow the directions.

1. Go south to the first turn.
2. Go east past the rabbit home.
3. Follow the path past mole's home.
4. Turn south and pass badger's art studio.
5. Go west past the skunks' home.
6. Turn south into the first doorway past the skunk home. This is the place for the special delivery!

Name _____

Toy Store Confusion

Felix has a huge delivery of toys for the toy store.

All the toys need to be put on the shelves, but he is not sure where to put them.

Draw a symbol in the key to match each word.

Draw each symbol on one of the boxes to show what is in the box.

This will make it easier for Felix to get the toys in the right places.

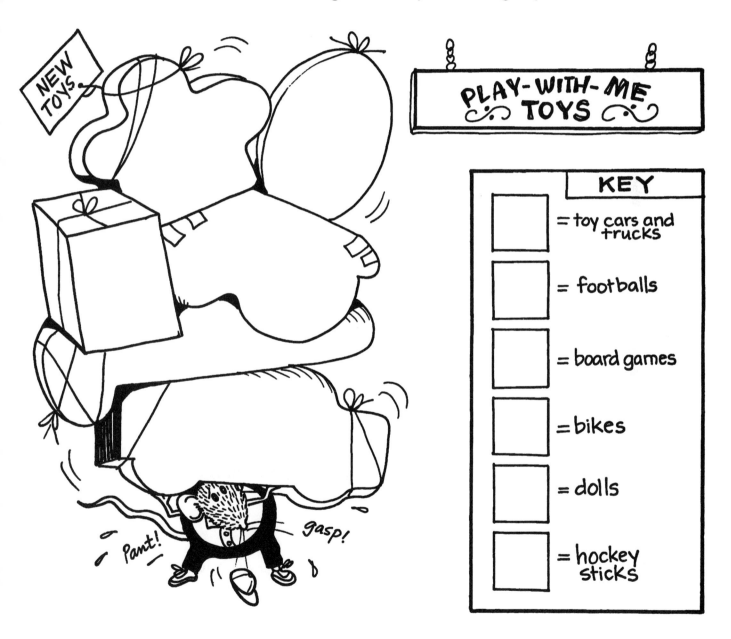

NEW TOYS

PLAY-WITH-ME TOYS

Pant!

gasp!

KEY

= toy cars and trucks

= footballs

= board games

= bikes

= dolls

= hockey sticks

Name _____

Mush! Mush! Rush!

Help Natuk make his deliveries—he's in a big hurry!
The map scale shows that 1 inch on the map equals 10 feet.
Use an inch ruler to measure from one dot to the other.

FISH CAKES

Fannie Freeze

The Icebergs

The Shivers

MMMM Fish cakes!

The Frostbites

Gabe Goosebump

Yahoo!

Hooray!

The Blizzards

1 inch = 10 feet
Scale

Ingrid Icicle

1. How far will Natuk travel from Fannie's to the Frostbites'? _____ feet

2. Whose home is 20 feet northeast of Gabe's? _____

3. How far is it from Ingrid's to the Shivers'? _____ feet

4. How far will he travel from Ingrid's to the Blizzards'? _____ feet

5. How far is it from the Blizzards' to Gabe's home? _____ feet

6. Who lives farther from the Icebergs, the Frostbites or Ingrid? _____

Name _____

The Picnic under the Picnic

The Bear family does not know that another picnic is being planned right beneath their feet!

The scale shows that every centimeter on the map equals 1 meter in the ant tunnels.

Use a centimeter ruler to connect the dots and measure the distances.

Scale

| 1 cm = 1 meter |

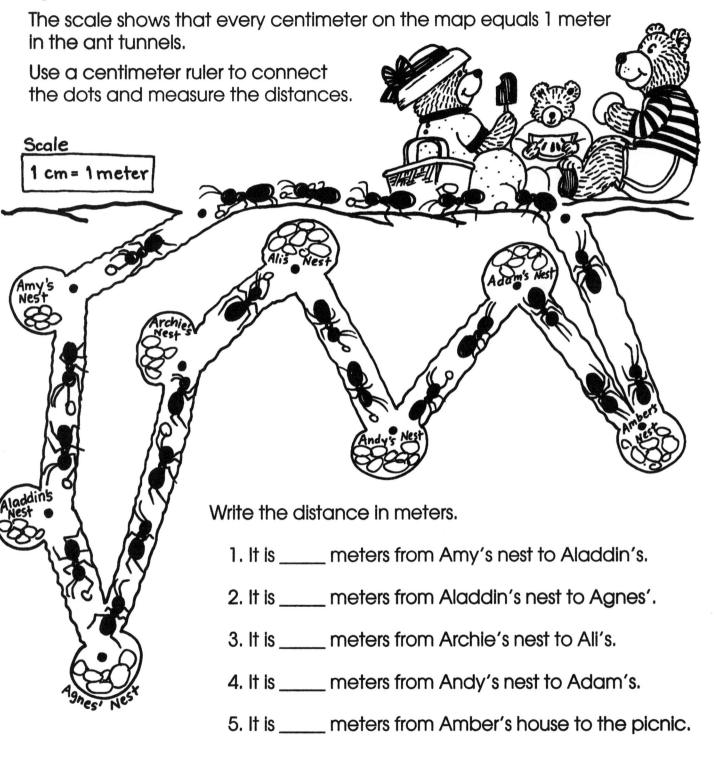

Write the distance in meters.

1. It is _____ meters from Amy's nest to Aladdin's.

2. It is _____ meters from Aladdin's nest to Agnes'.

3. It is _____ meters from Archie's nest to Ali's.

4. It is _____ meters from Andy's nest to Adam's.

5. It is _____ meters from Amber's house to the picnic.

Name _____

Island Hopping

Trader Doug the Sea Dog is always looking for a good trade.
He travels many miles trading things between the islands.
Use the scale on the map to find out how far he will travel today.
Use an inch ruler to measure the distance between the dots.

1. Doug took coconuts from Coconut Grove to Fish Lagoon. How far did he go? _____

2. He took a barrel of fish from Fish Lagoon to Flower Isle, _____ miles away.

3. He traded flowers for bananas on Banana Island—_____ miles from Flower Isle.

4. Next he went to Crab Atoll, _____ miles away, to trade bananas for crab cakes.

5. Then he headed for Home Island, _____ miles away from Crab Atoll, for a dinner of bananas and crab cakes.

Name _____

So Many Boxes!

Everything needed on Eagle Island has to be delivered by boat or helicopter.

Read the labels on the boxes.

Decide where each item needs to go on the map.

Draw a line from each box to the landform where the delivery will be made.

Fish Boat for Breezy Bay.

Motorboat for Lovely Lake.

Snowshoes for Mt. Tip-top.

Tractor for the field on the plain.

Gold pans for Nugget River.

Trees for Flat Plateau.

Log cabin for valley.

Lightbulbs for lighthouse on peninsula.

EAGLE ISLAND

Name _____

Lobsters for the World

Sam's Seafood Company delivers fresh lobsters from Boston to places all over the world. He needs to know a lot about the world in order to find all the right places.

Most maps of the world are flat.

But Sam knows that the Earth is not flat. It is round, like a ball.

So he uses a globe to help him with his travels.

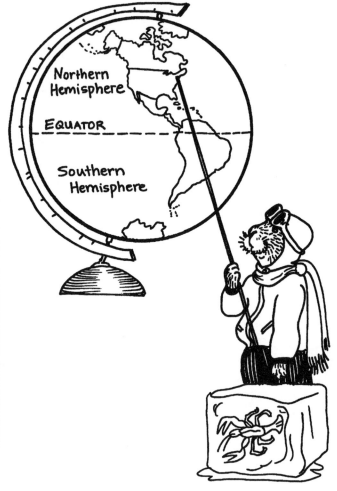

> A globe is a model of the Earth.
>
> The Earth is divided into 2 parts by an imaginary line called the equator.
>
> Each half of the Earth is called a hemisphere. This term means "half a ball."
>
> The equator divides the Earth into the Northern and Southern Hemispheres.

Help Sam get to know the globe!

1. Trace the equator with a red crayon.

2. Color the Northern Hemisphere yellow. Color the Southern Hemisphere blue.

3. Draw a purple flag on the North Pole. Draw a red flag at the South Pole.

4. What continent is entirely in the Northern Hemisphere?

_____

5. What continent is in both hemispheres?

_____

6. What continent is totally in the Southern Hemisphere?

_____

Name _____

Use with page 31.

Globe • Hemispheres

Lobsters for the World, cont.

Sam knows that the Earth is also divided into halves from the North Pole to the South Pole.

This makes the Eastern Hemisphere and the Western Hemisphere.

Sam has two maps that show him all four hemispheres.
1. Color the Western Hemisphere pink.
2. Color the Eastern Hemisphere yellow.

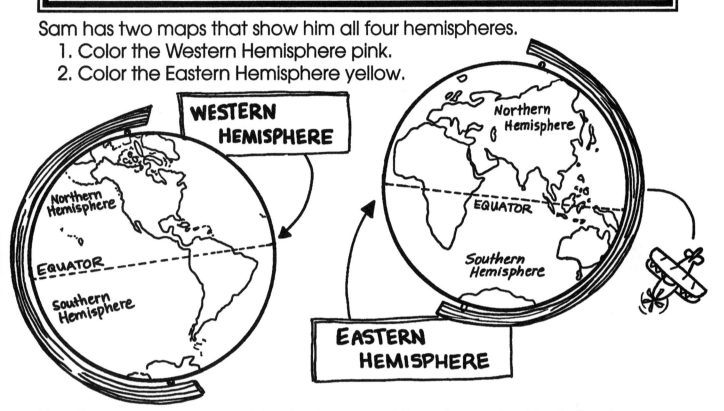

Use these maps to help him find some of the places for his deliveries.

Fill in the blanks to show what hemispheres each continent is in.

3. North America: Northern and _____ Hemispheres

4. South America: Southern, Western, and _____ Hemispheres

5. Australia: Southern and _____ Hemispheres

6. Africa: Eastern, Northern, and _____ Hemispheres

7. Asia: Northern, Southern, Western, and _____ Hemispheres

8. Europe: Eastern and _____ Hemispheres

9. Antarctica: Southern, _____, and _____ Hemispheres

Name _____

Use with page 30.

Globe • Hemispheres

Delivery Anywhere!

The U-Send-It Delivery Company will deliver to any place in the world. Here are just a few of the things they've been asked to deliver. Follow the directions to show the continents and oceans they've visited.

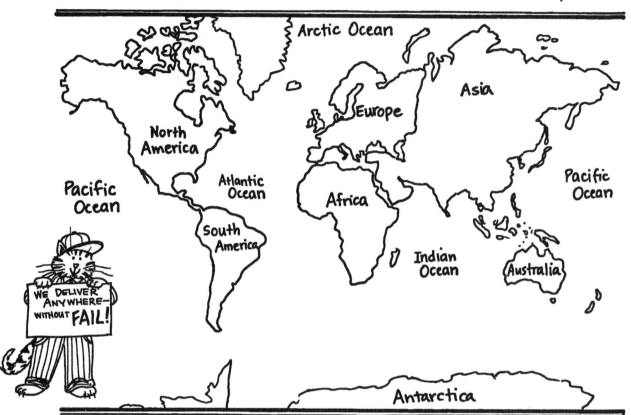

Follow the directions to color each place that receives a delivery. The name of the place for the delivery is in bold type.

U-Send-It Delivery delivers . . .
1. snowmobiles to **Antarctica**. .... Color it yellow.
2. ice cream to ocean liners in the
   Atlantic Ocean and **Pacific Ocean**. .... Color them green.
3. baby cribs to a day care center in **Europe**. .... Color it blue.
4. food to walruses in the **Arctic Ocean**. .... Color it purple.
5. a gorilla to return to its home in **Africa**. .... Color it red.
6. a wedding cake to a village in **South America**. ...... Color it brown.
7. fresh pineapple from Hawaii to **Asia**. .... Color it orange.
8. baby penguins to a zoo in **North America**. .... Color it black.
9. dinosaur bones to a museum in **Australia**. .... Color it red.
10. 20 scuba divers to the **Indian Ocean**. .... Color it green.

Name _____

Copyright ©1998 by Incentive Publications, Inc., Nashville, TN.
Basic Skills/Map Skills & Geography 2-3

Truckloads of Telephones

Tamara Talker has been hired to deliver telephones all over North America. Help her learn about the continent by following these directions.

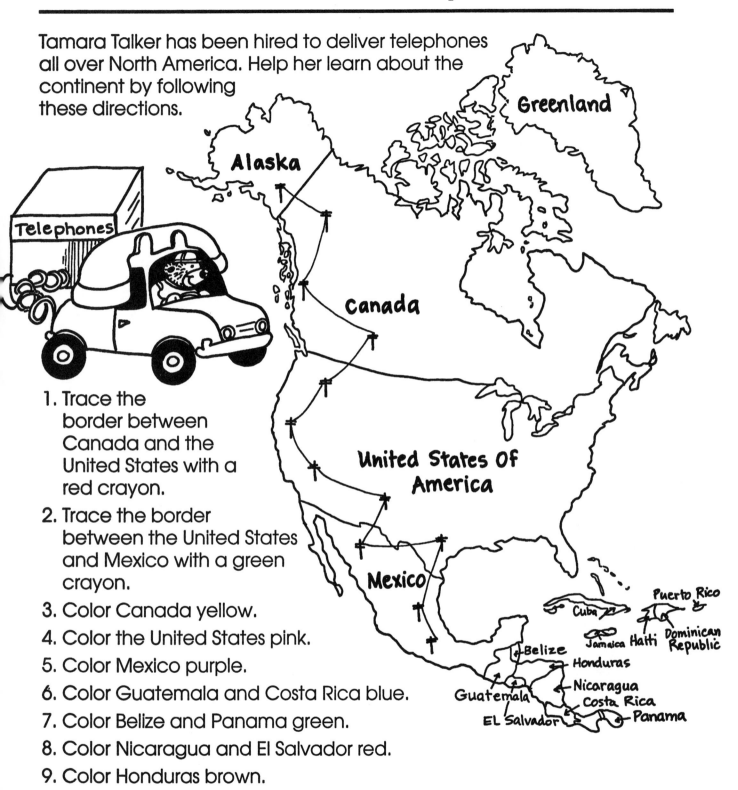

1. Trace the border between Canada and the United States with a red crayon.

2. Trace the border between the United States and Mexico with a green crayon.

3. Color Canada yellow.

4. Color the United States pink.

5. Color Mexico purple.

6. Color Guatemala and Costa Rica blue.

7. Color Belize and Panama green.

8. Color Nicaragua and El Salvador red.

9. Color Honduras brown.

10. Put a ★ on the place where you live.

Name _____

Mix-up at the Post Office

Oops! All the mail fell into a jumble.

Help the mail carriers find the right letters to put into their bags.

Hal Horse only delivers letters to states. Color his mail and bag red.

Gail Snail only delivers letters to cities. Color her mail and bag green.

Gilbert Gull only delivers letters to countries. Color his mail and bag blue.

Name _____

Cities, States, & Countries

Copyright ©1998 by Incentive Publications, Inc., Nashville, TN.
Basic Skills/Map Skills & Geography 2-3

Packages without Names

Charlotte has packages to deliver to different states in the United States.
Each package has a picture of the state but no name.
Can you recognize the states from their shapes?
Write the name of the state on each package. (See some clues below.)

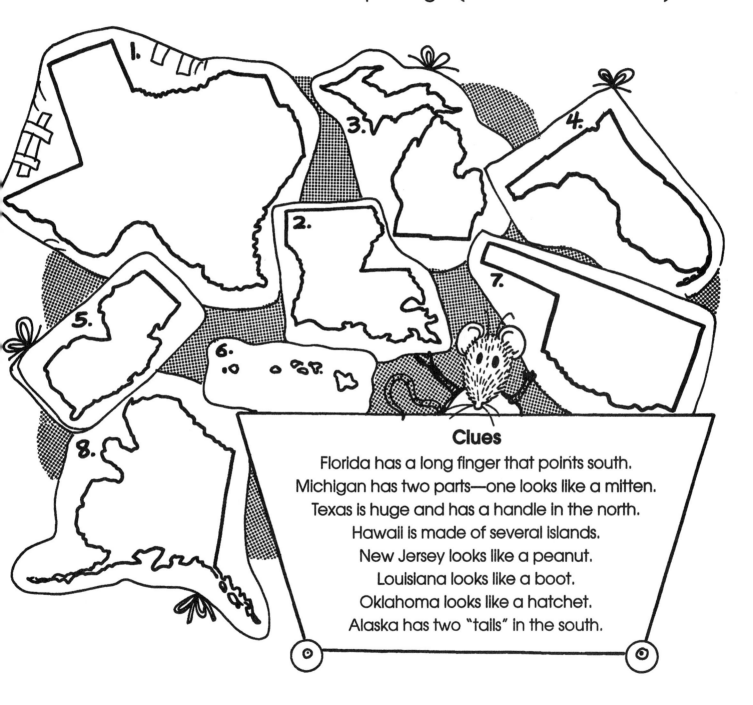

Clues

Florida has a long finger that points south.
Michigan has two parts—one looks like a mitten.
Texas is huge and has a handle in the north.
Hawaii is made of several islands.
New Jersey looks like a peanut.
Louisiana looks like a boot.
Oklahoma looks like a hatchet.
Alaska has two "tails" in the south.

Name _____

Rabbit Relatives

Grandma and Grandpa Rabbit send mail to grandchildren all over the United States.

They have a lot of letters and packages ready for the mailman.

Use the map on the next page to show where their mail will go.

Follow the color code to color each state correctly.

Color Code   Package

1. **Red** — a birthday card to little Roberta Floppy in St. Louis

2. **Green** — three boxes of carrot cookies to Phil in Philadelphia, Newt in New Orleans, and Bob in Boston

3. **Orange** — a model spaceship to Billie in the state where rockets are launched at Cape Canaveral

4. **Purple** — toys to triplets Sally, Sammy, and Sandy near the Grand Canyon

5. **Blue** — red sneakers to Ralph, who lives just west of the middle part of Lake Michigan

6. **Yellow** — balloons to Danny in the city of Dallas

7. **Pink** — a box of pretzels to Trina, who lives near the Golden Gate Bridge

8. **Brown** — valentines to Vicki and Val, who live near Great Salt Lake

9. **Red** — a rattle to newborn Willie, who lives near Mt. Rushmore

10. **Green** — skis to Adam in Seattle

11. **Blue** — brownies to little Millie in Niagara Falls

12. **Yellow** — a guitar to Gus in Nashville

13. **Purple** — mittens to Carrie and Connie in Connecticut

14. **Orange** — a kite to Henry in the state that touches Michigan and Lake Erie

Use with page 37.

U.S. Cities, States, & Landmarks

Basic Skills/Map Skills & Geography 2-3

The United States

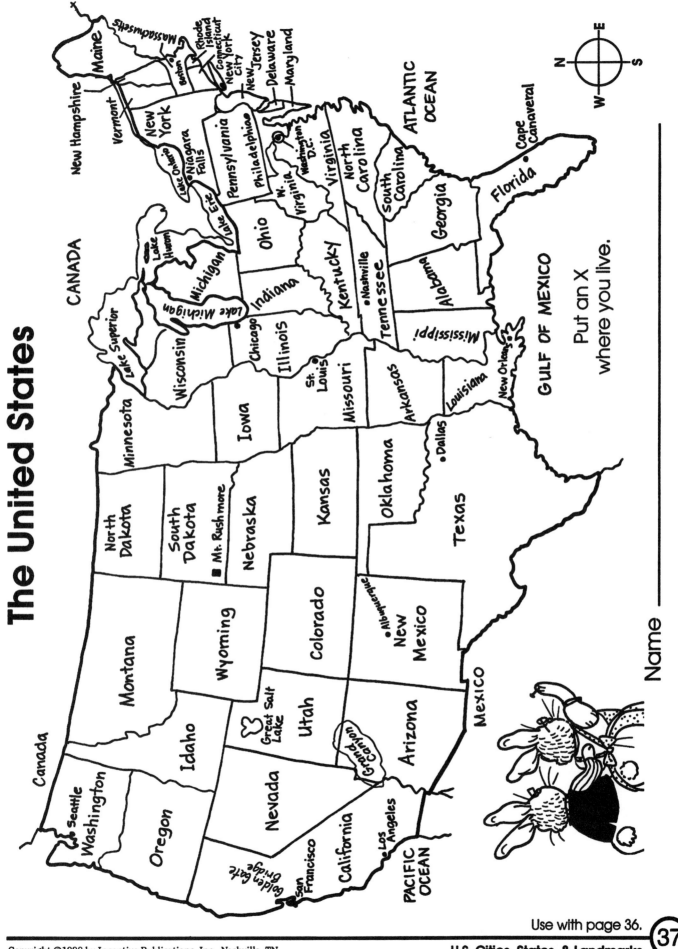

CANADA

ATLANTIC OCEAN

Maine
New Hampshire
Vermont
Massachusetts
Boston
Rhode Island
Connecticut
New York City
New Jersey
Delaware
Maryland

New York
Lake Ontario
Niagara Falls
Lake Erie
Pennsylvania
Philadelphia
W. Virginia
Washington, D.C.
Virginia
North Carolina
South Carolina
Georgia
Florida
Cape Canaveral

Ohio
Indiana
Kentucky
Nashville
Tennessee
Alabama

Lake Superior
Lake Huron
Lake Michigan
Michigan
Wisconsin
Chicago
Illinois
St. Louis
Missouri
Arkansas
Mississippi
Louisiana
New Orleans

Minnesota
Iowa

North Dakota
South Dakota
Mt. Rushmore
Nebraska
Kansas
Oklahoma
Dallas
Texas

Montana
Wyoming
Colorado
New Mexico
Albuquerque

Idaho
Great Salt Lake
Utah
Arizona
Grand Canyon

Washington
Seattle
Oregon
Nevada
California
Los Angeles
San Francisco
Golden Gate Bridge

Canada
Mexico

GULF OF MEXICO

PACIFIC OCEAN

Put an X where you live.

Name _____

U.S. Cities, States, & Landmarks

Lost Again!

Wrong-Way Walrus is lost again! Every time he goes out on a seaweed delivery, he ends up in the wrong body of water!

This time he popped up in the Gulf of Mexico instead of Boston Harbor.

He needs to check his map and hurry on his way before all the seaweed ice cream melts!

Use the map on the next page (page 39) to help you follow these directions and answer the questions.

1. Draw a sailboat in the ocean that touches the east coast of the U.S.

2. Color the Great Lakes green. How many are there? _____

3. Draw a whale in the Pacific Ocean.

4. Trace the Colorado River in red.

5. Trace the Missouri River in blue.

6. Trace the Mississippi River in green.

7. Trace the St. Lawrence River in yellow.

8. Draw a red arrow to Boston Harbor.

9. The X on the map shows where Wrong-Way Walrus is. What body of water is he in? _____

10. What river flows into the Pacific Ocean? _____

11. What river flows into the Mississippi from the east? _____

12. Name a city on the Missouri River. _____

13. Name a city on the Rio Grande River. _____

14. Name a city on the Gulf of Mexico. _____

15. What river flows along the border between the U.S. and Mexico? _____

16. Name a city on Lake Michigan. _____

Name _____

United States Bodies of Water

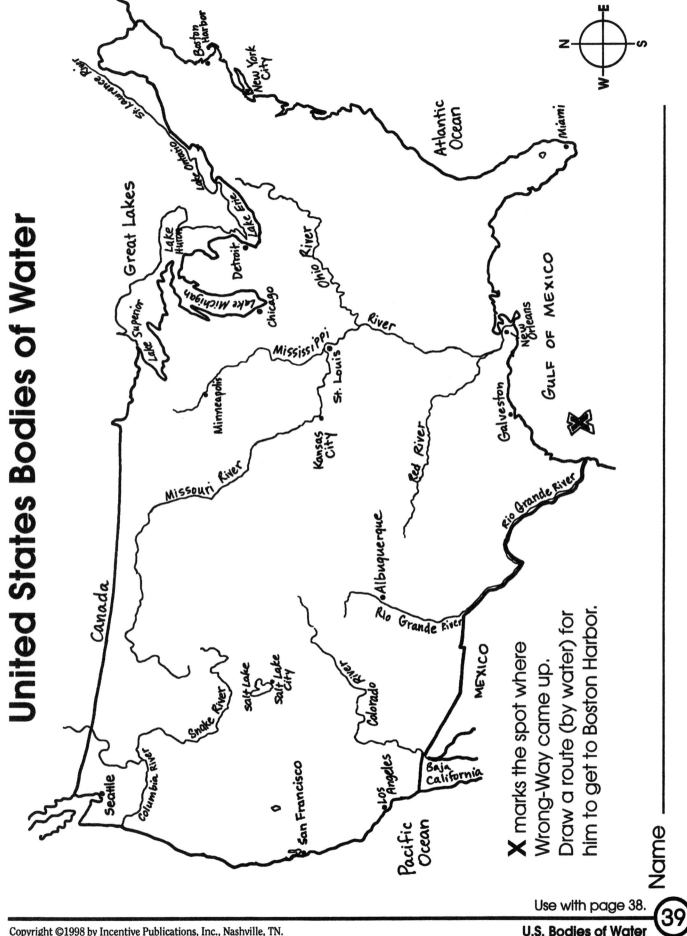

Great Lakes

St. Lawrence River

Boston Harbor

New York City

Lake Ontario

Lake Erie

Atlantic Ocean

Miami

Lake Superior

Lake Huron

Detroit

Lake Michigan

Chicago

Ohio River

River

Gulf of Mexico

New Orleans

Mississippi

Minneapolis

St. Louis

Kansas City

Missouri River

Galveston

Red River

Rio Grande River

Albuquerque

Rio Grande River

MEXICO

Canada

Salt Lake

Salt Lake City

Snake River

Colorado River

Los Angeles

Baja California

Seattle

Columbia River

San Francisco

Pacific Ocean

X marks the spot where Wrong-Way came up. Draw a route (by water) for him to get to Boston Harbor.

Name _____

The Lost Stagecoach

Death Valley Dan, the stagecoach driver, is lost!
He wonders if he will ever find his way to San Francisco with his bags of gold.

SOUTHWESTERN STATES

Redwood N.P.

Nevada

Great Salt Lake

Sacramento
Carson City

San Francisco

California

Yosemite N.P.

Great Basin

Utah

Provo
Salt Lake City

Colorado River

Boulder
Denver

Colorado

Pike's Peak

Death Valley N.P.

Las Vegas

Grand Canyon N.P.

Mesa Verde N.P.

Los Angeles

Pacific Ocean

Colorado River

Arizona

Rio Grande

Santa Fe

Albuquerque

San Diego

Baja

Phoenix

New Mexico

Mexico

Tuscon

Carlsbad

U.S. MINT U.S. MINT

Help Dan read his map.

1. What state touches the Pacific Ocean? _____

2. What state is south of Utah? _____

3. The Grand Canyon is in _____

4. San Francisco is in the state of _____

5. What state is east of Arizona? _____

6. Is Nevada east or west of California? _____

7. Which state is home to the Great Salt Lake? _____

8. Where is Pike's Peak? _____

9. Dan is now in the state east of Utah. Where is he? _____

10. How many national parks are shown? (Look for N.P.) _____

Name _____

Skeletons for the Schools

Gabe's job is to deliver science equipment to schools all over the United States.

The mileage chart tells him how far it is between cities.

Use the chart to help him find the distances.

Cities	Chicago	Boston	Denver	Nashville	Washington DC
Chicago	0	1015	1010	930	715
Los Angeles	2030	3025	1020	2030	2690
Miami	1370	1480	2080	900	1040
Seattle	2030	3090	1300	2430	2790
Tucson	1740	2630	890	1615	2280

Write the answers in the blanks.

1. Chicago to Seattle _____ miles

2. Tucson to Nashville _____ miles

3. Washington, DC, to Los Angeles _____ miles

4. Boston to Miami _____ miles

5. Miami to Washington, DC _____ miles

6. Denver to Tucson _____ miles

7. Which 2 cities are farthest apart? _____ and _____

8. Which 2 cities are closest together? _____ and _____

Name _____

Read a Chart

Fresh Air for Floats

Wanda has an unusual job! She pumps air into the floats at water parks.
She gets to try out the floats at every park, too!
Use the information chart to answer some questions about the parks
she visits.

Fresh Air

Putt Putt

Mobile

PARKS	Number of Water Slides	River	Open Year-Round	Overnight Camping	Snack Bar	Adults Allowed	Floats for Rent
Water Wonderworld	12	X	X		X	yes	X
Wild West Waterworks	7	X		X	X	yes	X
Dino-Land Park	0	X	X	X	X	no	
Water Thrills & Chills	10	X	X		X	yes	
Wet & Wild	14	X	X			no	X

Write the answers.

1. Can Wanda camp overnight at Wet & Wild?_____

2. Will Wanda have floats to fill at Dino-Land Park?_____

3. Can Wanda get a snack at Wild West Waterworks? _____

4. How many water slides are there at Water Thrills & Chills? _____

5. How many water slides are there at Water Wonderworld? _____

6. Is there a river at Wet & Wild? _____

7. Are adults allowed at Dino-Land? _____

8. Are adults allowed at Wild West? _____

9. Can Wanda camp overnight at Dino-Land? _____

10. Which park is not open year-round?_____

Name _____

Underwater Invitations

The Sea Bottom Ball is the biggest event of the year in the underwater world.
Oliver is delivering the invitations today.
Help Oliver get them to the right places.
Find the square on the grid that holds each undersea object.

For example: The starfish is in the square where row F and row 2 meet.
It is located at F, 2 on the grid.

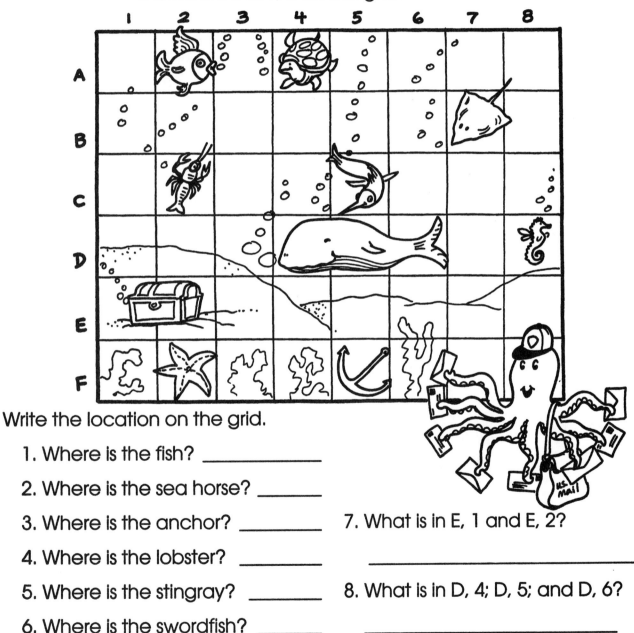

Write the location on the grid.

1. Where is the fish? _____

2. Where is the sea horse? _____

3. Where is the anchor? _____

4. Where is the lobster? _____

5. Where is the stingray? _____

6. Where is the swordfish? _____

7. What is in E, 1 and E, 2?

_____

8. What is in D, 4; D, 5; and D, 6?

_____

Color the picture.

Name _____

Read a Grid

Worms for the Garden

Gardener Gus is getting some nice, fat worms for his garden.
The worms make the soil good for growing healthy plants.
Help the worms find their way to the right spots in the garden.

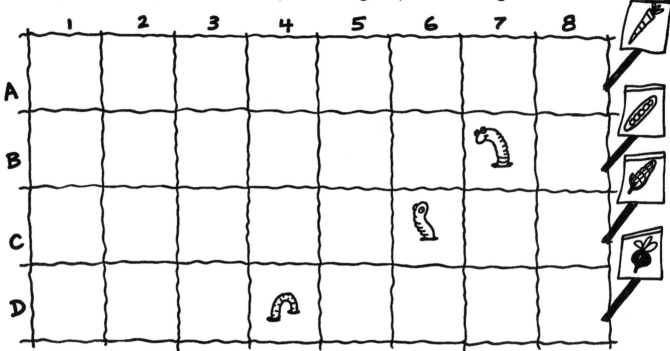

1. Some worms are already in the garden. Where are they?

   _____ , _____ , and _____

2. Draw a worm in A, 5.

3. Draw two worms in D, 7.

4. Draw a worm in B, 3.

5. Draw a worm in C, 2.

6. Draw a worm in B, 1.

7. Draw a worm in A, 4.

8. Draw a worm in D, 6.

9. Draw a worm whose head comes up in C, 6 and tail comes up in B, 8.

10. Draw two worms in the same hole in D, 2.

Name _____

Place Objects on a Grid

Delivery to Happy Camp

Bernie the Mail Bear has bug spray for the campers at Happy Camp. Finish the map that shows his trail through the campground.

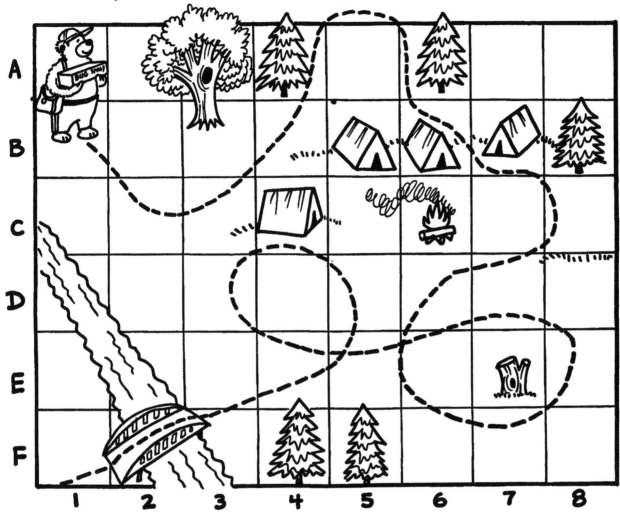

Draw these things on the map in the correct places.

1. Bernie hopped over a puddle ⟨⟩ at B, 2.
2. He went around a big tree 🌲 at A, 5.
3. Bernie put a package inside a tent ⟨△⟩ at C, 8.
4. Then he tripped over a rock ◯ in E, 6.
5. He bumped into a trash can 🗑 in E, 5.
6. He stopped to rest on a stump 🪵 in D, 4.
7. He crossed over a _____ in F, 2.

Name _____

Search for the Missing Flowers

The delivery bees are trying to gather honey to deliver to the beehive, but all the flowers are missing!

Follow the directions, and draw the flowers for the bees to find.

1. Draw 2 ❁ in D, 2.

2. Draw a ♄ in A, 2 and A, 7.

3. Draw a 🌀 in C, 1 and F, 7.

4. Draw a ✾ in D, 7 and F, 4.

5. Where is the fountain? _____

6. Where is the beehive? _____

7. Where is the sandbox? _____

8. Where is the slide? _____

Name _____

Flowers for the Queen

Beatrice's Flower Shop has a delivery of flowers for the queen bee.
Beatrice needs a grid to find the queen's chamber in the hive.
Use the grid and the map key to help Beatrice find her way around the
hive.

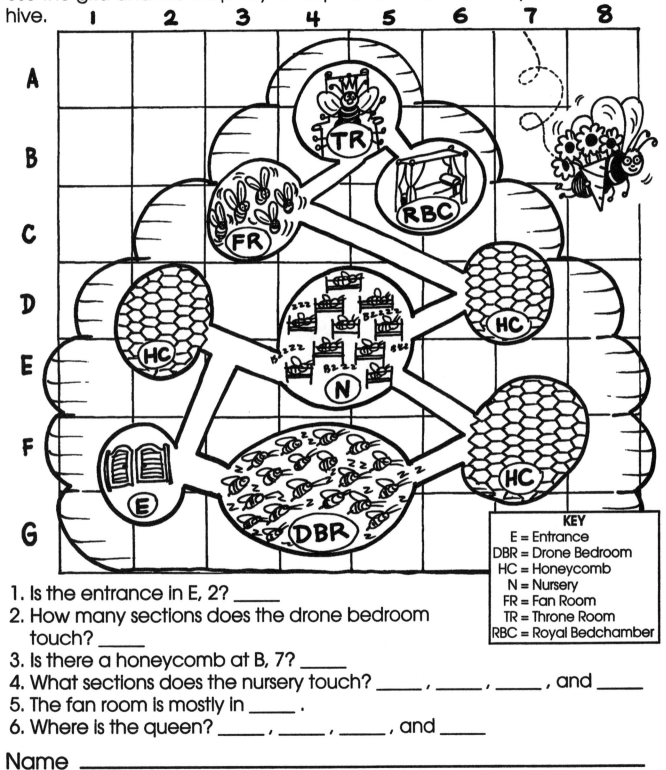

KEY
E = Entrance
DBR = Drone Bedroom
HC = Honeycomb
N = Nursery
FR = Fan Room
TR = Throne Room
RBC = Royal Bedchamber

1. Is the entrance in E, 2? _____
2. How many sections does the drone bedroom
   touch? _____
3. Is there a honeycomb at B, 7? _____
4. What sections does the nursery touch? _____ , _____ , _____ , and _____
5. The fan room is mostly in _____ .
6. Where is the queen? _____ , _____ , _____ , and _____

Name _____

Read a Grid & a Map Key

The Whereabouts of Bears

The hungry bears in Jellyshine Park are eating the campers' food.
Camper food is not good for bears, so the ranger has ordered
tons of berries.

The truck driver has a map of the bear population in Jellyshine Park.
Use the map to help her figure out where all the bears are.

1. Which section of the park will need the most berries? _____

2. What is the bear population in the Winter Caves Area? _____

3. Which has more bears: the campground or the Green Lakes District?

   _____

4. What is the bear population in the Canyon Cliffs Area? _____

5. Which area has the smallest bear population? _____

6. What is the bear population in the campground? _____

Name _____

Use a Population Map

Copyright ©1998 by INCENTIVE PUBLICATIONS, Inc., Nashville, TN.
Basic Skills/Map Skills & Geography 2-3

No Shooting Allowed

In Africa, dozens of photographers track endangered animals. They only shoot these animals with cameras!

Frannie Photog's map of Africa's endangered species helps her get to the right places.

AFRICA'S ENDANGERED SPECIES

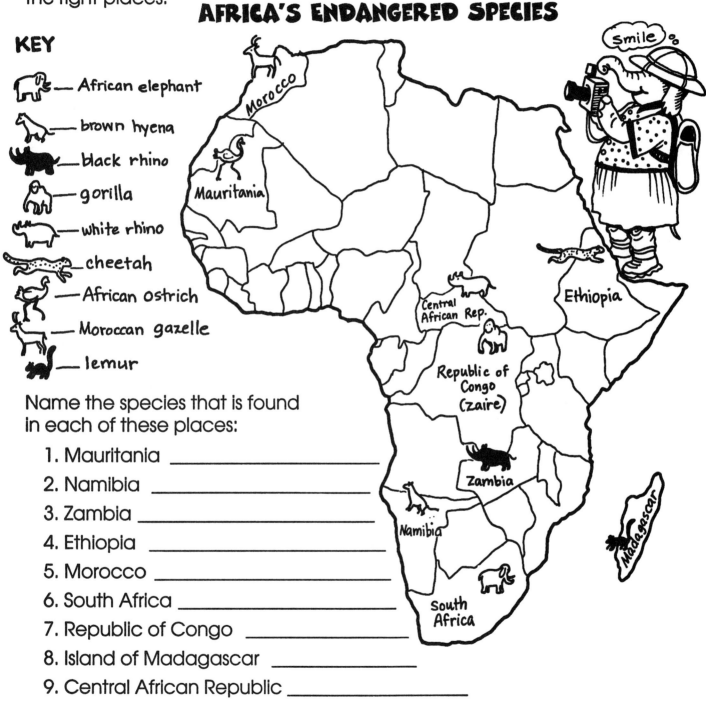

KEY

— African elephant

— brown hyena

— black rhino

— gorilla

— white rhino

— cheetah

— African ostrich

— Moroccan gazelle

— lemur

Name the species that is found in each of these places:

1. Mauritania _____

2. Namibia _____

3. Zambia _____

4. Ethiopia _____

5. Morocco _____

6. South Africa _____

7. Republic of Congo _____

8. Island of Madagascar _____

9. Central African Republic _____

Name _____

Use an Environmental Map

Bananas for the Neighbors

Juan grows bananas to be delivered all over South America!
His map shows some products produced in the countries he visits.

Write the name of a country that
shows these products on the map.

1. cattle and sheep

_____

2. coffee, fish,
   and oil

_____

3. cattle, oil,
   and coffee

_____

4. fish

_____

5. grain

_____

6. diamonds

_____

Write the answers.

7. Name a product
   produced in Columbia.

_____

8. Name a product
   produced in Bolivia. _____

9. Where is the most
   timber grown? _____

10. In what country does
    Juan grow bananas? _____

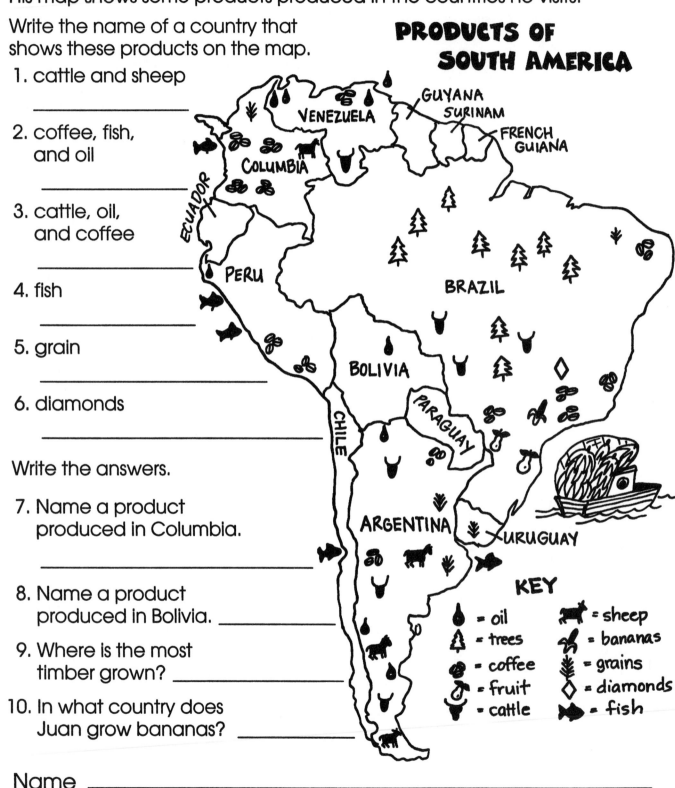

PRODUCTS OF SOUTH AMERICA

KEY

● = oil   🐑 = sheep
🌲 = trees   ⚘ = bananas
☕ = coffee   ⚹ = grains
🍐 = fruit   ◇ = diamonds
🐂 = cattle   🐟 = fish

Name _____

In All Kinds of Weather

Bouncing Barb the mail carrier never fails to deliver the mail on time!
She carries a weather map so she will be prepared for all kinds of weather.
Use her weather map to answer the questions about Australia's weather.

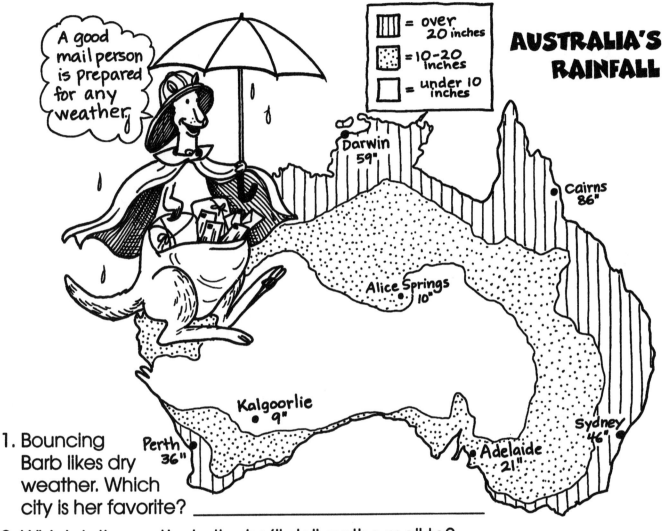

1. Bouncing Barb likes dry weather. Which city is her favorite? _____

2. Which is the wettest city she'll deliver the mail to? _____

3. Which gets more rain, southern or eastern Australia? _____

4. Which part of the country is the driest, the middle or the coastline? _____

5. Can they expect more rain in Kalgoorlie or Perth? _____

6. What is the rainfall each year in Sydney? _____

7. What rainfall is expected each year in Darwin? _____

Name _____

Use a Rainfall Map

On the Road Again

Freddy Fox has to drive from Cocoaville to Sweet Town to deliver chocolate chips to the Crispy Cookies Company.

It's a long way, but Freddy has a good road map.

Read his map to find the answers to the questions.

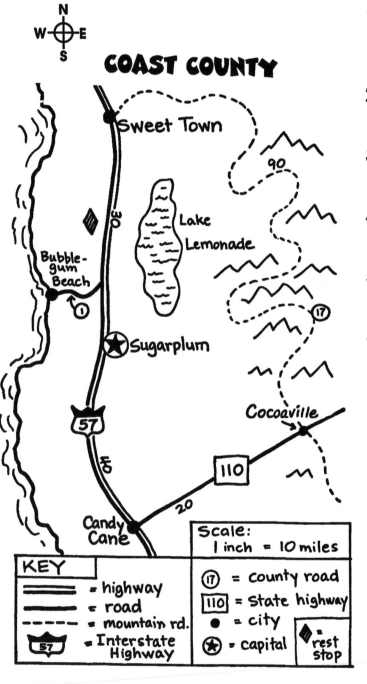

1. What capital city is shown?

   _____

2. On the scale,
   1 inch = _____ miles.

3. Candy Cane is on
   Interstate Number _____ .

4. Does Route 110 go
   through Cocoaville? _____

5. Is Lake Lemonade
   east of Interstate 57? _____

6. The county road goes
   from Cocoaville to

   _____ .

7. How far is it from Cocoaville
   to Candy Cane?

   _____ miles

8. What route goes
   from Interstate 57 to
   Bubblegum Beach?

   _____

Name _____

A Puzzling Delivery

Miss Cross, the puzzle champion, has waited all day for the newspaper with the new crossword puzzle in it.

Finally it has arrived!

Today's puzzle uses map words.

Help her solve it with the words in the **Word Box.**

WORD BOX

poles	key
equator	bay
continent	road
west	states
peninsula	ocean
scale	sea
hemisphere	island

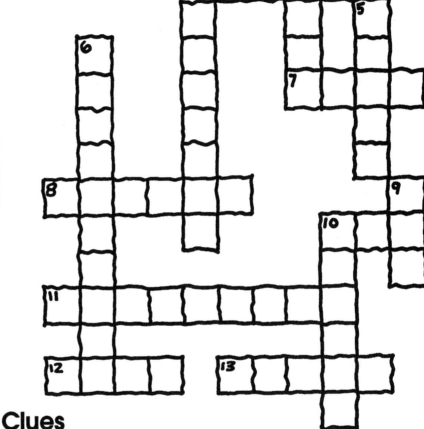

At last!

Clues

Across

1. part of a map that shows symbols
4. largest body of water in the world
7. way to travel between cities
8. land surrounded by water
10. another name for ocean
11. land surrounded on 3 sides by water
12. the opposite of east
13. The _____ are at the top and bottom of Earth.

Down

2. imaginary line around center of Earth
3. large body of land
5. tells distances on a map
6. half of the Earth
9. area of water sheltered by land
10. The United States has 50 of these.

Name _____

Where Are the Elephants?

Homer has a load of hay for the circus elephants.
Complete the directions on the next page to help Homer find the elephant compound.

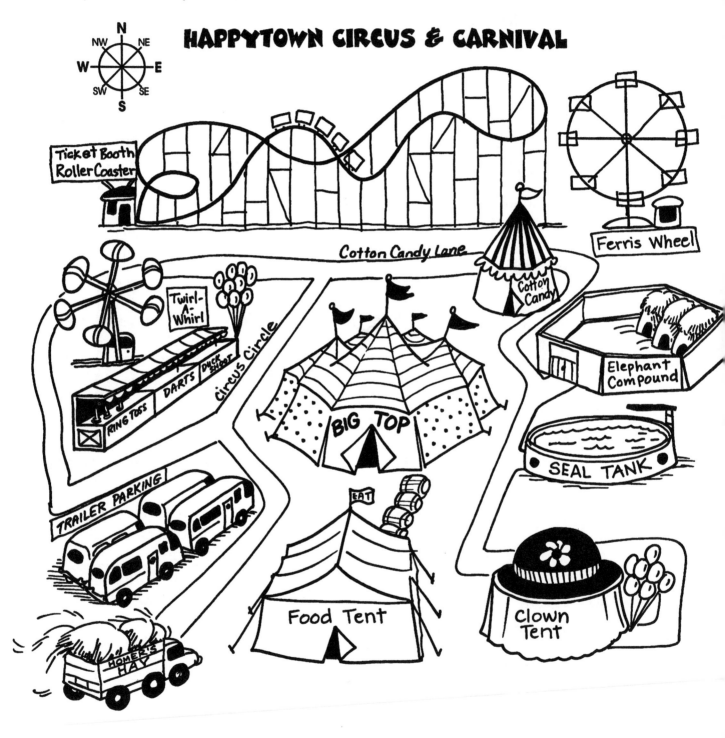

HAPPYTOWN CIRCUS & CARNIVAL

Ticket Booth
Roller Coaster

Ferris Wheel

Cotton Candy Lane

Cotton Candy

Twirl-A-Whirl

Circus Circle

RING TOSS   DARTS   DUCK SHOOT

Elephant Compound

BIG TOP

SEAL TANK

TRAILER PARKING

EAT

HOMER'S HAY

Food Tent

Clown Tent

Name _____

Use with page 55.

Write Directions

Use the map on page 54 and the compass to complete the directions below.

Homer travels _____ until the road turns _____ . When the road splits at Circus Circle, he turns right and travels _____ . At Cotton Candy Lane, he turns right again and travels _____ . Oops! There is a big curve in the road. Homer turns _____ and then _____ .

Whew! He finally reaches the elephant compound.

Delivery for Elmo.

pant

Good! Just in time for lunch!

Follow the directions and answer the questions.

1. Color the ride that is northeast from the Big Top Tent.

2. Color the tent that is southeast from the Big Top Tent.

3. Color the booth that is in the northwest corner.

4. What is southwest of the Big Top? _____

5. Will Homer find elephants west of the Big Top? _____

6. Is the Roller Coaster north of the Big Top? _____

7. Is the Ring Toss west of the Big Top? _____

8. Is the Food Tent north of the Roller Coaster? _____

9. Is the Clown Tent farther east than the hay truck? _____

10. Can you buy cotton candy to the southeast of the Food Tent? _____

Name _____

Write Directions

Where Are the Stamps?

Edward, the mail bear, is waiting at the door for Gilda's letter. Gilda is madly searching her room for a stamp to put on the letter. Where are the stamps?

Make a map of Gilda's room.

Include these things on the map:

- **a title**
- **a scale**
- **symbols**
- **a key**
- **a place for the stamps**

Mail?

What a silly goose I am! Where did I put those stamps?

Title:

Scale:

Key

Name _____

Map & Geography Skills Test

Look for these parts of the map. Write the letter for each one next to its name.

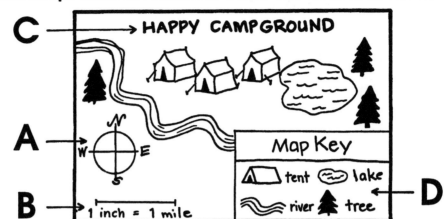

____ 1. title

____ 2. key

____ 3. scale

____ 4. compass

5. Which one is the map that correctly shows the things on the table? Circle the letter of the correct map.

A   **B**

Use the map to answer the questions. Write N, S, E, or W in each blank.

6. The town is _____ of the airport.

7. The river is _____ of the town.

8. The lake is _____ of the river.

9. The town is _____ of the lake.

10. The airport is _____ of the lake.

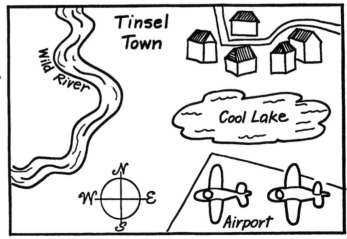

Name _____

Use the map key to match the words to the symbols.
Write the correct letter next to the word for each symbol.

_____ 11. river

_____ 12. house

_____ 13. railroad track

_____ 14. airport

_____ 15. forest

_____ 16. bridge

Write the correct letter from the map to match each landform name.

_____ 17. island

_____ 18. mountain

_____ 19. lake

_____ 20. plateau

_____ 21. peninsula

_____ 22. valley

23. Circle the state that is in the western part of the United States.

New York   Florida   Michigan   California

24. Circle the state that is in the southern part of the United States.

California   Colorado   Florida   Maine

25. Circle the state that is in the eastern part of the United States.

Texas   New Jersey   Arizona   California

Name _____

Use the maps to find the hemispheres for these continents and oceans.
Write E for Eastern, W for Western, and B for both.

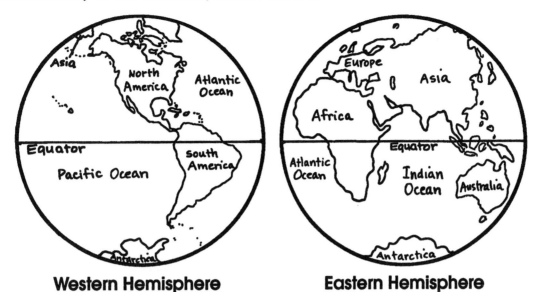

Western Hemisphere   **Eastern Hemisphere**

_____ 26. Africa   _____ 27. The Atlantic Ocean   _____ 29. Europe

                        _____ 28. South America   _____ 30. Asia

31. Circle the word that is NOT a continent.
    Asia   Africa   Texas   North America   Australia

32. Circle the word that is NOT a city.
    Dallas   Chicago   San Francisco   Europe   Seattle

33. Circle the word that is NOT a state.
    Iowa   Pennsylvania   Kansas   Boston   Georgia

34. Circle the word that is NOT a country.
    France   India   Chicago   Mexico   Canada

35. Circle the country that is on the northern border of the United States.
    Alaska   Germany   Canada   Montana   Mexico

36. Circle the country that is a neighbor to the south of the United States.
    Japan   Mexico   Canada   China   Russia

Name _____

Map & Geography Skills Test

Use the grid to help you find things on the picnic table.

_____ 37. Where is the 🍎 ?

_____ 38. Where is the 🐜 ?

_____ 39. Where is the 🧁 ?

_____ 40. Where is the 🥪 ?

	1	2	3	4
A	grapes			
B		ant		apple
C	drink		cupcake	
D		sandwich		

Use the map to answer the questions.

_____ 41. How many rabbits are in the north area of the park?

_____ 42. What area has the least rabbits?

_____ 43. What area has 40 rabbits?

Rabbit Population in Animal Park

North Area

West Area

South Area

🐰 = 10 rabbits

Write the word that matches the definition.

_____ 44. tells distances on a map   west

_____ 45. land surrounded by water   continent

_____ 46. opposite of east   ocean

_____ 47. a large body of land   scale

_____ 48. a large body of water   title

_____ 49. tells what a map is about   compass

_____ 50. tells directions on a map   island

Name _____

Answer Key

Skills Test

1. C	26. E
2. D	27. B
3. B	28. W
4. A	29. E
5. A	30. B
6. N	31. Texas
7. W	32. Europe
8. E	33. Boston
9. N	34. Chicago
10. S	35. Canada
11. C	36. Mexico
12. A	37. B, 4
13. D	38. B, 2
14. E	39. C, 3
15. F	40. D, 2
16. B	41. 50
17. E	42. South
18. B	43. West
19. A	44. scale
20. D	45. island
21. F	46. west
22. C	47. continent
23. California	48. ocean
24. Florida	49. title
25. New Jersey	50. compass

Skills Exercises

pages 10–11

Addresses will vary.
Planet: Earth
Galaxy: Milky Way
Check to see that student answers are correct.

pages 12–13

The correct map is # 1.
Check to see that the map includes some representation of all items in their correct places.

page 14

Check to see that student has colored the map correctly.

page 15

1. #2	3. #1
2. #3	4. #2

Carl needs map #3 to find Maple Street.

page 16

The correct cave is the one in the far southeast corner (lower right).

page 17

1. N
2. NW
3. N
4. S
5. NW
6. Pelican's
7. Crocodile's
8. Turtle's
9. no
10. yes
11. Turtle's

page 18

Check to see that student has followed the directions accurately.

page 19

1. egg—brown
2. turtle shell—green
3. hive—yellow
4. tree—orange
5. cactus—green box
6. Fido's house & straw nest—brown with circle

Milk is delivered to the turtle shell, Fido's house, and the straw nest.

pages 20–21

1. yes
2–3. Look to see that students have traced and colored these correctly.
4. yes
5. no
6. yes
7. library
8. Bridge
9. Downtown
10. 3
11. workers at the Pencil Factory

page 22

See that student draws the correct candy in each space.

page 23

1. Cactus, Sagebrush, or Locoville
2-4. See that student has colored these features accurately.
5. 2
6. 2
7. twice
8. Dry Desert State Park

page 24

The delivery is for the rat family.

page 25

Check to see that student has drawn a reasonable symbol for each toy named.

page 26

1. 10
2. the Icebergs
3. 30
4. 15
5. 30
6. Ingrid

page 27

1. 6
2. 4
3. 4
4. 5
5. 6

page 28

1. 30 miles
2. 20
3. 40
4. 20
5. 30

page 29

Connect these boxes with a line to these landforms.
fish boat—bay
motorboat—lake
snowshoes—mountain
tractor—plain
gold pans—river
trees—plateau
log cabin—valley
lightbulbs—lighthouse

pages 30—31

1-3. Check to see that student has traced, colored, and drawn pictures according to the instructions.
4. North America
5. South America
6. Antarctica
1-2. Check to see that student has colored the hemispheres accurately.
3. Western
4. Northern
5. Eastern
6. Southern
7. Eastern
8. Northern
9. Eastern and Western

page 32

Check to see that student has colored each area accurately.

page 33

Check to see that student has traced or colored each area accurately.

page 34

RED—Hal Horse—California, Ohio, Colorado, New Hampshire, Montana, Alabama, Pennsylvania
GREEN—Gail Snail—San Francisco, Albuquerque, Dallas, Albany, New Orleans, Boston, Nashville
BLUE—Gilbert Gull—United States, Peru, Japan, India, Great Britain, Australia

page 35

1. Texas
2. Louisiana
3. Michigan
4. Florida
5. New Jersey
6. Hawaii
7. Oklahoma
8. Alaska

pages 36-37

1. Red—Missouri
2. Green—Pennsylvania, Louisiana, Massachusetts
3. Orange—Florida
4. Purple—Arizona
5. Blue—Wisconsin
6. Yellow—Texas
7. Pink—California
8. Brown—Utah
9. Red—South Dakota
10. Green—Washington
11. Blue—New York
12. Yellow—Tennessee
13. Purple—Connecticut
14. Orange—Ohio

pages 38-39

1, 3-8. See that student has drawn and traced things accurately.
2. 5
9. Gulf of Mexico
10. Columbia
11. Ohio
12. Kansas City
13. Albuquerque
14. Galveston
15. Rio Grande
16. Chicago

Check to see that student has drawn a route by water from the Gulf of Mexico to Boston Harbor.

page 40

1. California
2. Arizona
3. Arizona
4. California
5. New Mexico
6. east
7. Utah
8. Colorado
9. Colorado
10. 5

page 41

1. 2030
2. 1615
3. 2690
4. 1480

5. 1040
6. 890
7. Seattle and Boston
8. Chicago and Washington, DC

page 42
1. no
2. no
3. yes
4. 10
5. 12
6. yes
7. no
8. yes
9. yes
10. Wild West Waterworks

page 43
1. A, 2
2. D, 8
3. F, 5
4. C, 2
5. B, 7
6. C, 5
7. treasure chest
8. whale

page 44
1. D, 4; C, 6; and B, 7
2-10. Check to see that student has drawn worms in correct spots.

page 45
1-6. Check to see that student has drawn items in correct spots.
7. bridge

page 46
1-4. Check to see that student has drawn flowers in correct spots.
5. B, 3
6. E, 4
7. E, 2
8. C, 2

page 47
1. no
2. 6

3. no
4. D, 4; D, 5; E, 4; E, 5
5. C, 3
6. A, 4; A, 5; B, 4; B, 5

page 48
1. Green Lakes District
2. 70
3. Green Lakes District
4. 50
5. Big Trees Area
6. 40

page 49
1. African ostrich
2. brown hyena
3. black rhino
4. cheetah
5. Moroccan gazelle
6. African elephant
7. gorilla
8. lemur
9. white rhino

page 50
1. Argentina
2. Peru or Argentina
3. Venezuela or Argentina
4. Chile, Peru, Columbia, Uruguay, or Argentina
5. Uruguay, Argentina, Brazil, or Columbia
6. Brazil
7. grain, coffee, sheep, or fish
8. oil
9. Brazil
10. Brazil

page 51
1. Kalgoorlie
2. Cairns
3. eastern
4. middle
5. Perth
6. 46 inches a year
7. 59 inches a year

page 52
1. Sugarplum
2. 10 miles
3. 57
4. yes

5. yes
6. Sweet Town
7. 20
8. Route 1

page 53
Across
1. key
4. ocean
7. road
8. island
10. sea
11. peninsula
12. west
13. poles
Down
2. equator
3. continent
5. scale
6. hemisphere
9. bay
10. states

pages 54–55
Directions:
NE, NW, NE, E, SW, SE
1-6. Check to see that Ferris wheel, clown tent, and ticket booth are colored.
4. trailer parking or hay truck
5. no
6. yes
7. yes
8. no
9. yes
10. no

page 56
Check to see that student has included title, key, scale, symbols, and a place for the stamps to be lost.

Answer Key